CVX

PROPULSION SYSTEM DECISION

Industrial Base Implications of Nuclear and Non-Nuclear Options

John SCHANK • John BIRKLER • Eiichi KAMIYA
Edward KEATING • Michael MATTOCK
Malcolm MacKINNON • Denis RUSHWORTH

Prepared for the United States Navy

NATIONAL DEFENSE RESEARCH INSTITUTE

RAND

Approved for public release; distribution unlimited

PREFACE

The U.S. Navy is currently in the process of designing its next-generation aircraft carrier, termed CVX. One technological advance for the CVX, compared with the current *Nimitz* class of carriers, is a new propulsion system. To help in deciding between a nuclear and non-nuclear propulsion system for CVX, the Navy asked RAND to determine the effects of the CVX propulsion choice on the industrial bases supporting nuclear propulsion systems and those supporting conventional propulsion systems.

RAND began the research in April 1998 and, because of the timeframe of the CVX propulsion system decision, provided the results of the analyses to the CVX program office in August 1998. Shortly thereafter, the decision was made that the CVX class would use nuclear propulsion. The research findings described in this document helped influence that decision. This document offers the briefing presented to the program office along with accompanying annotation.

The research was based on understanding the various propulsion system options being considered and contacts with suppliers and integrators of key system components to establish how a CVX decision would affect their company and their industry, as well as other Navy programs.

This research and its documentation should be of interest to the Office of the Secretary of Defense (OSD) and Navy policymakers and planners who must face the propulsion system decisions for future Navy ship programs. It may also interest industrial decisionmakers involved in construction and supply of Navy ships and parts.

The research reported here was carried out in the Acquisition and Technology Policy Center of the National Defense Research Institute, RAND's federally funded research and development center (FFRDC) supporting the Office of the Secretary of Defense, the Joint Staff, the unified commands, and the defense agencies.

CONTENTS

Preface iii
Summary vii
Acknowledgments xi
Acronyms xiii
The Navy Is Transitioning to a New Class of
  Aircraft Carriers 1
Appendix
A. PRIME MOVER ATTRIBUTES 55
B. ENGINE EXHAUST AIR POLLUTANT EMISSIONS 57
C. ARRANGEMENT CONSIDERATIONS FOR THE AIR
   INTAKES AND EXHAUSTS FOR FOSSIL-FUELED
   AIRCRAFT CARRIERS 65
Bibliography 69

SUMMARY

CVX 78, the first ship in the new class of Navy aircraft carriers, will include a new propulsion system, a new electrical distribution system, and other modifications to incorporate the latest version of onboard electronic subsystems. Propulsion system options considered to replace the current nuclear system (the A4W) used on the *Nimitz*-class carriers include the design and development of a new nuclear reactor and several conventional (i.e., non-nuclear) systems—gas turbine, diesel, and oil-fired steam.

A decision regarding the prime mover portion of the overall propulsion system needed to be made by the end of fiscal year 1998 to meet the planned production schedule.[1] The timeliness of the decision was important. If a nuclear option was chosen, the new reactor needed to be designed and developed and the nuclear components (some of which have production lead times of five years or more) needed to be built. Since the reactor plant in a nuclear powered ship is one of the first components needed during the construction phase, any delay in the delivery of the components to the shipyard would delay the overall construction schedule of the ship.

---

[1] A ship's propulsion system has three major components: the prime mover, the drive system, and the propulsor. The prime mover converts the thermal energy of fuel (fossil or nuclear) into rotational kinetic energy, either directly, as in the case of diesel engines or gas turbines, or by creating steam to drive steam turbines. The output of the prime mover is then translated (and combined) by the drive system to the propulsor, usually a propeller with either fixed or controllable pitch. The Navy has used a variety of prime movers in recent decades, depending on the mission and capability of the ship.

Conventional propulsion systems are standard on commercial ships and for naval combatants, amphibious ships, and auxiliaries. Nevertheless, choosing a non-nuclear propulsion system for CVX 78 would involve a significant and unique design effort to integrate the propulsion system into the overall ship. For example, a gas turbine prime mover will require extensive intake and exhaust ducting that must be incorporated into the overall existing *Nimitz*-class hull form and structure. Additionally, some non-nuclear components (e.g., controllable pitch propellers and ship service turbine generators) have design and production lead times of more than eight years.

Because of the importance of the propulsion system decision for CVX, the Navy initiated a number of studies to understand the cost and performance trade-offs of nuclear versus non-nuclear propulsion for aircraft carriers. In addition, the CVX program office of the Naval Sea Systems Command (NAVSEA (PMS-378)) asked RAND to identify and quantify the effects on the industrial base resulting from the CVX prime mover decision. RAND began the research in April 1998 and, because of the timeframe of the CVX propulsion system decision, provided the results of the analyses to the CVX program office in August 1998.

RAND's analysis addressed the following questions. If the CVX propulsion were conventional (e.g., diesel or gas turbine), what would the effects be on the industrial base supporting conventional propulsion systems and the one supporting nuclear propulsion systems? Alternatively, if the CVX propulsion were nuclear, what would the effects be on these industrial bases? This document is organized around the analysis and findings in response to these questions.

Current suppliers of key components with an understanding of the various propulsion system options being considered were contacted to establish how a CVX propulsion system decision would affect their company and their industry, as well as other Navy programs.

Our analysis showed that neither a nuclear CVX nor a non-nuclear CVX would affect the conventional propulsion industrial base. The manufacturers in that area have robust markets in other Navy ships and/or in the commercial sector; the presence or absence of demand for a conventionally powered CVX would scarcely be felt.

However, the nuclear industrial base may be affected by the demand for either a nuclear or non-nuclear CVX. If CVX were conventional, the cost of components for other Navy nuclear programs would increase. The cost of the heavy equipment and cores for the construction of nuclear submarines and the midlife refuelings of carriers and submarines would increase by approximately $20 million to $35 million per year, or 5 to 7 percent of the cost of the nuclear components, depending on the program and the year. If CVX were nuclear, there is a potential schedule problem with the delivery of the heavy equipment components, suggesting that the CVX propulsion system decision must be made soon, and if nuclear, the reconstitution of production capability closely managed.

ACKNOWLEDGMENTS

This study was conducted over a four-month time period and greatly benefited from the assistance of many people in the Navy, at RAND, and elsewhere.

Within the Naval Sea System Command, CAPT Tal Manvel, then Program Executive Officer (PEO), Aircraft Carriers, and Brian Persons, his deputy, provided tremendous support and encouragement. Richard Williams, Robert Murphy, and William Schmitt of the Nuclear Power Directorate opened doors within their respective organizations and at contractor facilities. They went out of their way to be helpful and did the legwork that made our field trips possible. The management and engineers at BWX Technologies (BWXT) Incorporated, Nuclear Equipment Division, A McDermott Company, were gracious hosts and shared their experiences, extensive knowledge, and data with us. All their insights, suggestions, and cooperation were essential to completing this research on time.

Newport News Shipbuilding (NNS) hosted meetings with NNS technical staff and generously provided us with valuable information and insights into the ship construction impacts related to the nuclear/non-nuclear prime mover issue. GE Marine Engines and Wärtsilä Diesel Engines, with on-sight briefings and tours of their facilities, provided us with insights into ship installation issues, and a view of current industry capabilities. Gibbs & Cox was always available, via phone and e-mail, to answer those uniquely ship-engineering questions and to provide supporting documentation. It was the cooperation of all that allowed us to quickly come to understand the nature of the multiple options and appreciate their implications on the industry.

At RAND, Ron Hess provided a thorough and thoughtful technical review. Gordon Lee's advice on organization of the briefing is much appreciated.

Of course, we alone are responsible for any errors.

ACRONYMS

AoA	Analysis of Alternatives
AP	Advanced procurement
bhp	Brake horsepower
BWXT	BWX Technologies
BWXT/NED	BWXT, Nuclear Equipment Division
CPP	Controllable pitch propeller
CODAG	Combination diesel and gas turbine
CODOG	Combination diesel or gas turbine
CVX	Next generation aircraft carrier
DD	Destroyer
DDG	Destroyer, guided missile
EMALS	Electro-Magnetic Aircraft Launch System
EPA	Environmental Protection Agency
FBM	Fleet ballistic missile
FPP	Fixed pitch propeller
FFRDC	Federally funded research and development center
GAO	Government Accounting Office
GE	General Electric
gr/kw-hr	Grams per kilowatt-hour
hp	Horsepower
ICCLS	Internal Combustion Catapult Launch System
ICR	Inter-cooled recuperated
IDA	Institute for Defense Analysis
IMO	International Maritime Organization
IN	Indiana
KBHP	Thousands of brake horsepower
KHP	Thousands of shaft horsepower
LPD	Amphibious transport, docks

LTD	Limited
MA	Massachusetts
MAN	MAN Aktiengesellschaft
MIT	Massachusetts Institute of Technology
MR	Mentoring ratio
MW	Megawatts
NAVSEA	Naval Sea Systems Command
NDRI	National Defense Research Institute
NNS	Newport News Shipbuilding
NSSN	New class of attack submarine
OH	Ohio
OSD	Office of the Secretary of Defense
RCOH	Refueling complex overhauls
rpm	Revolutions per minute
SCR	Selective catalytic reduction
shp	Shaft horsepower
SSBN	Ballistic missile submarine, nuclear powered
TADA	Total available dock area
UK	United Kingdom
US	United States
USCG	United States Coast Guard
VA	Virginia
W/O	Without
WI	Wisconsin

> **The Navy Is Transitioning to a New Class of Aircraft Carriers**
>
> CVX 78, as planned, will include
> - A new propulsion system
> - A new electrical distribution system
> - Modifications to reduce total ownership cost
>
> A decision on the propulsion system's prime mover needs to be made soon
> - Nuclear components require long lead times
> - Conventional systems will require significant testing and integration into the ship
>
> NDRI   RAND

The U.S. Navy is in the process of constructing the last two ships of the *Nimitz* class of aircraft carriers. Newport News Shipbuilding (NNS) is in the middle of the construction program for CVN 76 (*Reagan*) with a scheduled delivery date of 2002. The contract award and construction start of the last of the *Nimitz* class of carriers, CVN 77, will commence in 2001 with a scheduled delivery to the Navy in 2008.

In addition to completing the *Nimitz* class of carriers, the Navy is beginning the design and development of the next-generation aircraft carrier, designated CVX. Because of the high cost of designing a new carrier and because of competing demands within its budget, the Navy has undertaken an approach that evolves the CVX design from the basic *Nimitz*-class design over several ships starting with CVN 77. CVX will possess new technologies and subsystems designed to either increase the performance of the ship or reduce its total ownership cost.

The first ship in the new class of aircraft carriers, the CVX 78, is expected to include a new propulsion system, electrical distribution system, and other modifications that incorporate the latest version of onboard electronic subsystems and provide warfighting improvements and life-cycle cost reductions. Propulsion system options being considered for CVX 78 include the design and development of a new nuclear reactor that would replace the current A4W systems used on the *Nimitz*-class carriers as well as several gas turbine, diesel, and oil-fired steam systems.

A decision regarding the prime mover portion of the overall propulsion system needed to be made by the end of fiscal year 1998 to meet the planned production schedule. The timeliness of the decision was important. If a nuclear option were chosen, the new reactor must be designed and developed and the nuclear components, some of which have production lead times of five years or more, must be built. Since the reactor plant in a nuclear powered ship is one of the first components needed during the construction phase, any delay in the delivery of the components to the shipyard would delay the overall construction of the ship.

Although conventional propulsion systems are widely used on commercial ships and for naval combatants, amphibious ships and auxiliaries, choosing a non-nuclear propulsion system for CVX 78 would still involve significant design efforts to integrate the propulsion system into the overall ship. For example, a gas turbine prime mover requires extensive intake and exhaust ducting that must be incorporated into the overall existing *Nimitz*-class hull form and structure. Additionally, like the nuclear components, some non-nuclear components (e.g., controllable pitch propellers and ship service turbine generators) have design and production lead times of over eight years.

> **RAND's Mission**
>
> **Research Objective**
> - Identify *industrial base implications* resulting from CVX prime mover decision
>
> **Research time frame**
> - April–August 1998
>
> **Research Caveat**
> - We did *not* address operational performance or total ownership costs
>
> NDRI   RAND

The propulsion system decision for CVX has motivated the Navy to undertake a number of study efforts to understand the cost and performance trade-offs of nuclear versus non-nuclear propulsion for aircraft carriers. Based on our previous research of the submarine and carrier industrial bases,[1] especially that part of the research that focused on the nuclear vendors that support those industrial bases, the CVX program office of the Naval Sea Systems Command (NAVSEA (PMS-378)) asked RAND to identify and quantify the industrial base implications resulting from the CVX prime mover decision.[2]

For our analysis, the industrial base includes the people, equipment, facilities, and other resources and processes of the prime contractors and their vendors that develop and produce either nuclear or conventional propulsion systems. We also examine potential effects at

---

[1] See Birkler et al., 1994, and Birkler et al., 1998.

[2] The overall propulsion system is composed of a number of components: the prime mover (nuclear reactor or gas turbine engines), the drive mechanism that translates and combines the output of the prime movers, and the propulsor (shafts and propellers).

Newport News Shipbuilding, the sole shipyard that currently builds aircraft carriers. We have not considered any possible effects on the Navy laboratories or other organizations (such as the Engineering Directorate of the Naval Sea Systems Command) that develop or support conventional or nuclear propulsion systems on naval ships.

RAND began the research in April 1998 and, because of the timeframe of the CVX propulsion system decision, provided the results of the analyses to the CVX program office in August 1998. The RAND effort focused solely on the implications for the industrial bases. Other organizations examined the operational performance and total ownership cost issues surrounding nuclear versus non-nuclear propulsion for aircraft carriers.[3]

---

[3]See, for example, Whiteway and Vance, 1998.

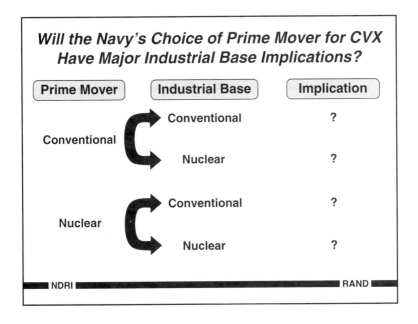

The analysis addressed four similar questions based on the choice of prime mover (either conventional or nuclear) and the resulting effect on the two different industrial bases—the industrial base supporting conventional propulsion systems and the one supporting nuclear propulsion systems. These four questions, or cases, are shown in the above chart.[4] For example, one of the cases studied was the effect on the conventional industrial base if the decision is made to use a conventional prime mover for CVX.

This documented briefing presents the analysis and findings for each of the four cases presented in this chart.

---

[4]During the course of the research, the CVX Program Office posed two additional questions concerning the pollution emissions from conventional propulsion systems and the placement within an aircraft carrier of the air intakes and exhausts needed for a gas turbine propulsion system. These two issues are addressed in Appendixes B and C, respectively.

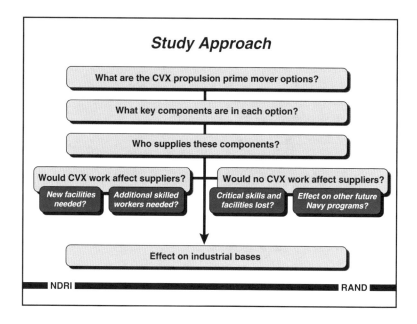

This chart describes the study approach utilized for the research effort.

Step one was to gain an understanding of the various propulsion system options being considered as well as their key components and basic operations.[5] Suppliers of these components were identified and attempts were made to determine if they could be future suppliers. The shrinking base of Navy contracts has caused restructuring within the industry and, for a variety of reasons, many historic suppliers are no longer available.

The next step was to contact dominant suppliers and establish what the effects of a CVX propulsion system decision would be for their particular industry in general, and specifically for their company. Effects of the CVX decision on other Navy programs were also studied.

We were interested in a variety of issues for each supplier. On one hand, we wanted to know if CVX propulsion system work would in-

---

[5]Propulsion system options for the CVX were defined during the CVX Analysis of Alternatives (AoA). See AoA, 1998.

crease their current workload to the point where additional production resources—facilities, manufacturing equipment, skilled workers—would be needed. On the other hand, we wanted to know if the loss of CVX work would adversely affect suppliers. That is, would the absence of CVX workload lead to the loss of critical skills or production resources needed to maintain capability for other naval programs. In both cases, we wanted to quantify the cost of increases or decreases in the workload of the nuclear and conventional propulsion system industrial bases on other naval programs.

How We Studied the Questions

Interacted with
- NAVSEA
- Wärtsilä NSD
- GE Evendale
- Alstom-Cegelec
- NNS
- Coltec Industries
- GE Lynn
- Northrop Grumman
- Gibbs & Cox
- BWX Technologies

Built Analytical Tools
Model to determine feasible workforce expansion

Analyzed Data
Current and future nuclear programs
NNS workforce requirements
Cost of NSSN and RCOH ship sets

NDRI — RAND

We interacted with key conventional and nuclear propulsion system vendors and organizations. The Nuclear Propulsion Directorate of NAVSEA (SEA 08) provided valuable insights into the status of its vendors and identified those that might be adversely affected if CVX were non-nuclear. The primary nuclear vendor that raised some concern was BWX Technologies (BWXT). It is the sole supplier of heavy-equipment components for both submarine and carrier propulsion sets. We also worked closely with NNS to understand the effect on its nuclear workforce of the CVX propulsion system decision. For the past 35 years, it has been the sole shipyard building U.S. aircraft carriers.

Suppliers of conventional propulsion system components that shared data and information with us included Wärtsilä NSD, a Finnish company that is one of the world leaders in diesel engine production; Coltec, a U.S. company supplying the diesel engines for the LPD-17 (amphibious transport, docks) program; and General Electric (GE) in Evendale, OH, the world leader in turbine engine production. Also, GE in Lynn, MA, the supplier of mechanical drive systems for U.S. naval ships, and Alstom-Cegelec, a multinational organization that provides electric drive systems for ships, shared their thoughts and expertise with us. Gibbs & Cox, a leading U.S. naval architecture firm, and Northrop Grumman, the manufacturer of the ship service turbine generators for submarines and aircraft carriers also provided information and assistance during the research effort.

These organizations provided valuable insights and supplied data that helped us analyze the effects of alternative prime mover options for CVX. For the analysis of the nuclear industrial base, these data included projections of current and future workloads associated with various nuclear programs for both new construction and reactor refuelings. The organizations also supplied workforce requirements associated with different nuclear workload demands and the costs of nuclear components for both submarine new construction and carrier refuelings.

We built an analytical tool to help understand the time required to rebuild and expand a workforce under a variety of assumptions.

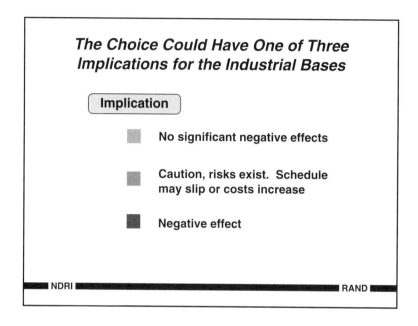

To analyze each of the four cases posed during the research, we used three different symbols to indicate the degree to which problems could exist. A green box suggests there are no significant negative effects on any segment of the industrial base due to the decision on the CVX prime mover. An orange box indicates there are some cautions, either in meeting required schedules or in increased costs to other naval programs. A red box indicates a negative effect on a segment of the industrial base or a potential "show stopper" associated with the CVX prime mover decision.

As we progress through the briefing, this scheme is shown in the lower right corner of the charts to indicate our assessment of potential industrial base problems, using the box that best suggests the risk involved for each of the four cases being evaluated.

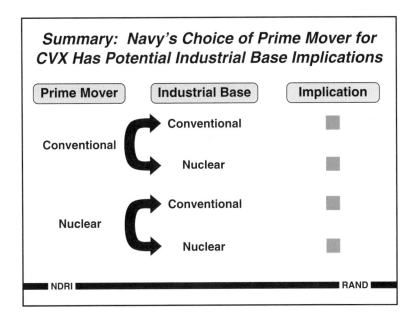

This chart summarizes the significant findings of our research. Our analysis suggests there are no significant negative effects on the conventional propulsion system industrial base regardless of whether the CVX uses a conventional or nuclear prime mover. However, there are some cautions regarding the nuclear industrial base or other Navy programs no matter which prime mover is chosen for the CVX. The remainder of this document describes the analysis that led to these conclusions.

> **We Considered Two Viable Conventional Prime Mover Classes**
>
> - **Diesel**
> - **Gas Turbine**

The CVX AoA identified three classes of conventional prime mover options—(1) oil-fired steam, (2) diesel, and (3) gas turbine.[6] The oil-fired steam option was not considered in this study because of the diminishing presence of such systems on any new Navy designs, as well as discussions with NAVSEA's Power System Group (SEA 03Z) on the likelihood of the option.[7] Very few U.S. Navy ships built in the last 30 years have used oil-fired steam as the prime mover; most surface combatant, amphibious, and auxiliary ships have used either diesel or gas turbine engines. The commercial ship industry has also moved almost completely away from steam-powered ships. Although steam is fairly efficient at cruise speeds and provides high output power, steam propulsion plants are large and heavy. They also have high manning requirements and provide rather limited endurance.

---

[6]Prime mover attributes are summarized in Appendix A.

[7]Although oil-fired steam systems were not considered for the CVX prime mover as part of this study, selecting either a diesel or gas turbine system for the CVX would still require some type of steam boiler system to generate steam for the aircraft catapults until the Electro-Magnetic Aircraft Launch System (EMALS) becomes operational.

The diesel and gas turbine prime movers include (in addition to the basic power plant) ancillary equipment such as cooling, intake, and exhaust subsystems; shock and noise reduction subsystems; and monitoring and control subsystems. Typically, these various subsystems are packaged together by the diesel or turbine manufacturer into a complete "system" that is delivered to the shipyard. The shipyard provides the foundation for the propulsion system plus connects the intake and exhaust subsystems to the ducting built during ship construction. The procedure for a nuclear propulsion system is different: SEA 08 contracts for numerous specific components of the nuclear reactor, and these components are delivered to the shipyard, which functions as the propulsion system integrator.

While the study guidelines were limited to the prime mover options, the drive options were also reviewed sufficiently to determine if they had an effect on, or were affected by, the prime mover decision. In conversations with NAVSEA and the prime mover and drive suppliers, it became clear that such effects were minimal and not critical to prime mover industrial base issues.

Also, no industrial base capability currently exists in the United States for electric drive systems. Therefore, the analysis of the conventional options focused on diesel and gas turbine systems using mechanical drive systems.[8]

---

[8]During the course of the analysis, performance concerns eliminated diesel system prime movers as a viable option. Nevertheless, we have included our findings on the diesel industrial base.

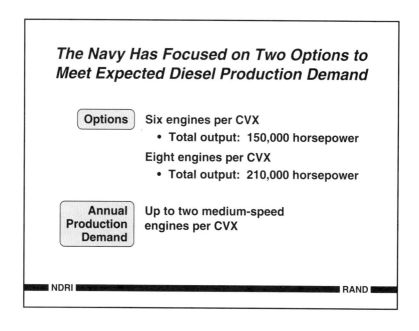

There are three general types of diesel engines—slow speed (less than 200 rpm), medium speed (between 200 and 1,000 rpm), and high speed (greater than 1,000 rpm). Although capable of producing over 70,000 horsepower, slow-speed diesels are very tall, which affects ship configurations, and do not have the engine speed and power characteristics suitable for carrier operations. High-speed engines do not generate sufficient horsepower to meet the CVX requirement. Therefore, medium-speed diesels are the candidates for CVX. Medium-speed diesel engines have a sizable presence in the U.S. Navy, primarily on amphibious and auxiliary ships, in addition to being widely used by commercial ships.

Diesel engines have a number of advantages and disadvantages compared with other prime mover options for CVX. They are highly fuel-efficient at almost all power loads and can use lower-cost fuels. They also permit lower manning than that required for current steam or nuclear systems. However, they generate higher pollution levels than most other types of engines, although existing engines are being modified to reduce pollutant emissions.[9] The exhaust air may also cause problems with landing aircraft on the flight deck and raises concerns of potential aircraft and external ship structure corrosion.

The CVX AoA identified two basic diesel engine configurations for the CVX. The first option used six engines per ship with a total output of 150,000 shaft horsepower.[10] The second option required eight engines per carrier with a total output of 210,000 shaft horsepower. The AoA designated the Wärtsilä 64 family of medium-speed engines. The diesel engine options would place an average annual demand of up to two medium-speed engines on the diesel industrial base.

---

[9] See Appendix B for a discussion of emission concerns for diesel and gas turbine engines.

[10] Shaft horsepower is the power required at the propeller shafts. Brake horsepower is the power output of the engine and will be referred to as "horsepower" in this report. The difference between brake horsepower and shaft horsepower is the losses in the drive mechanism.

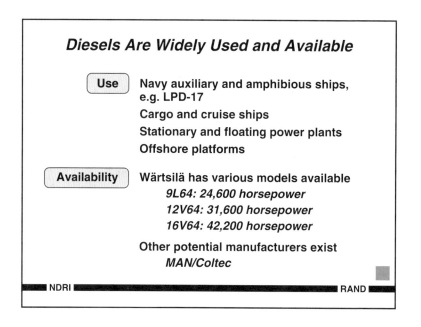

In addition to commercial cargo and cruise ships, diesel engines are widely used on naval non-combatant ships. They are also widely used on stationary and floating power plants as well as offshore oil platforms.

The Navy would find extensive available industrial capacity if it chooses a diesel-powered CVX. Based in Finland, but with production plants around the world, Wärtsilä NSD has a family of diesel engine models that meet CVX requirements. Its "64" family of diesels can be built with various numbers of cylinders to provide different horsepower outputs. The largest three engines in the family are shown on the chart. A 12-cylinder version has been delivered for a stationary power application. A 7-cylinder engine has been sold for installation on a commercial ship. Wärtsilä NSD has leased a facility in Mount Vernon, IN, in hopes of acquiring U.S. Navy business. With the loss of the LPD-17 propulsion system to Coltec, the Indiana facility has excess capacity that could easily handle CVX demands.

Fairbanks Morse Engine is a division of Coltec Industries located in Beloit, WI, and is a licensee of S.E.M.T. Pielstick, a French diesel engine manufacturer that is 50 percent owned by MAN Aktiengesellschaft, a German company. MAN's diesel engine production and deliveries are comparable in size to Wärtsilä NSD. Coltec currently builds diesel engines for the Navy Sealift and LPD-17 programs. Coltec has a joint development and license agreement with MAN for low-emission, dual-fuel engines.

Because of the wide use and availability of diesel engines, we believe there is little or no risk to the diesel production base if the decision is made to use diesel engines on the CVX class of carriers. Therefore, we assign a green box.

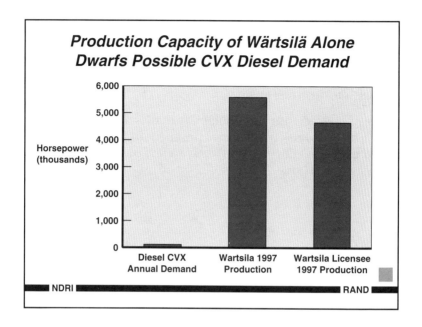

This chart compares Wärtsilä NSD's and its licensees' 1997 production to the Navy's prospective demand for CVX diesel engines as measured in thousands of horsepower (KHP). A diesel CVX would require production of about 70 KHP per year, which is a small fraction of Wärtsilä NSD's or MAN's capacity.

In 1997, Wärtsilä NSD delivered engines totaling 5,560 KHP, and its licensees supplied an additional 4,620 KHP. Furthermore, MAN's annual report states that in 1996–1997, MAN B&W Diesel Group and its licensees attained a 60 percent share of the world's market for low-speed diesel engines to power ships, and a 33.5 percent share (600 engines) for medium-speed ship engines.

Because the average annual demand from a diesel CVX is so small compared with average demands from the commercial shipbuilding industry, we believe there is no effect on the diesel industrial base resulting from a decision to use diesel engines on the CVX. Therefore, a green box is assigned.

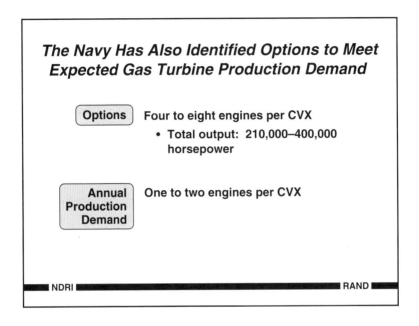

With its lightweight, fast response, and agility, the gas turbine has been found to be ideal for the smaller surface combatants of the U.S. and foreign navies. Gas turbines also provide high fuel efficiencies if operated at full power and require less maintenance and manning than either oil-fired steam or nuclear systems. However, gas turbines also have some disadvantages. They require large intake and exhaust ducts to provide air to the turbine and to vent the exhausts outside of the ship structure.[11] This exhaust air generates a high infrared signature. The exhaust air may also cause problems with landing aircraft on the flight deck and raises concern about potential aircraft and external ship structure corrosion. Gas turbines also have low fuel efficiency if operated at partial power.

The CVX AoA identified a number of propulsion system options that use gas turbines as the prime mover. These options, designated the General Electric marine family of engines, were the LM2500, LM2500+, and LM6000. Depending on the size of the engine, various gas turbine options specified between four and eight engines per

---

[11]See Appendix C for a discussion of intake and exhaust ducts.

ship with total shaft horsepower ranging from 210,000 to 400,000.[12] The gas turbine options would place an average annual demand of from one to two engines per year on the gas turbine industrial base.

---

[12]Additionally, two to four gas turbine generators would be required to support ship service electrical loads.

Gas Turbines Are Widely Used and Available

Use
Commercial ships, high-speed boats
Navy surface combatants, e.g., DDG-51 class
Stationary power plants
Jet airplanes, helicopters

Availability
General Electric has various models available
LM2500: 31,200 horsepower
LM2500+: 39,000 horsepower
LM6000: 60,150 horsepower

Other potential manufacturers exist
Allison/Rolls Royce
Northrop Grumman
Pratt & Whitney

NDRI — RAND

As previously mentioned, gas turbines are widely used on naval and commercial ships. Most surface combatants built in the last 20 years for U.S. and foreign navies have used gas turbines for propulsion. High-speed ferry boats and pleasure craft also use gas turbines for propulsion, and Royal Caribbean Cruise Line has chosen gas turbines for its new cruise ships.

Gas turbine engines traditionally have been developed for aircraft and later modified for stationary and marine use. Smaller industrial and marine gas turbines are derivatives of military engines, but larger versions that would be appropriate for the CVX are derivatives of commercial airline engines. For example, the GE LM2500 engine that is used on the U.S. Navy's DDG-51 class of surface combatants is a derivative of the GE CF-6 used on the DC-10 and C-5 aircraft. Similarly, the LM6000 is a descendant of the CF6-80C2 used on the Boeing 747 aircraft.

Aircraft engine designs undergo significant modifications such as removing compressor stages and adding turbine stages and mechanical power takeoffs to adapt them for stationary industrial applications. Stationary engines are further modified for marine use through shock mounting and salt-air protection. The technology for such modifications is well understood.

This chart shows the horsepower output of the three GE gas turbine engines listed in the CVX AoA. These engine outputs do not include any intake and exhaust losses that might be incurred because of unique ducting that would be needed for the CVX. Newport News Shipbuilding has conducted design exercises with alternative propulsion systems and has analyzed the duct losses for several possible CVX installations. It rates the output of the GE engines as approximately 28,000 horsepower for the LM2500; 33,000 horsepower for the LM2500+; and 40,000 horsepower for the LM6000.[13]

As mentioned, the LM2500 is used widely in marine applications, including the DDG-51 class of surface combatants. The LM2500+ has been modified for commercial marine use and will be used on the new line of Royal Caribbean Cruise Line ships. The LM6000 is available for stationary industrial applications but has not been modified for marine or military use. However, the LM6000 is a candidate for the U.S. Navy's new DD-21 class of surface combatants. A number of other companies, including Rolls-Royce, Northrop Grumman, and Pratt & Whitney, also manufacture gas turbine engines.

Since turbine engines have wide use in marine applications and are readily available, we believe there would be no negative effect on the gas engine industrial base. Depending on the specific manufacturer and model choices for the CVX, there may be some development and testing required to modify the engine for naval marine applications. But such modifications have occurred in the past and are well understood by the gas turbine industrial base. Therefore, we assign a green box.

---

[13] Newport News Shipbuilding, 1998.

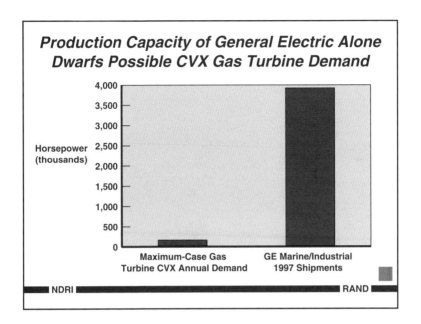

As with diesel engines, current gas turbine industry capacity dwarfs possible CVX demands.

Depending on the shaft horsepower of the specific option chosen and based on the options listed in the CVX AoA, each CVX carrier would require from four to eight engines. For the LM6000, one CVX every four years would translate into an average of two LM6000s per year, or roughly 120,000 horsepower worth of engines in the steady state. By contrast, GE's marine and industrial engine division shipped over 3.9 million horsepower worth of gas turbine engines in 1997.

This figure does not include the massive aircraft gas turbine engine industry nor any firm other than General Electric. For example, according to company sources, Rolls-Royce, which produces aircraft, industrial, and marine gas turbines, delivered 1,300 gas turbine engines in 1997. Pratt & Whitney, which has produced about 1,000 industrial and marine gas turbines, claims to power more than half of the Western-built large commercial jet transports currently in service.

Again, because of the large manufacturing capacity for turbine engines, especially compared with the average annual demands from a CVX, we believe there would be no negative effect on the industrial base and assign a green box.

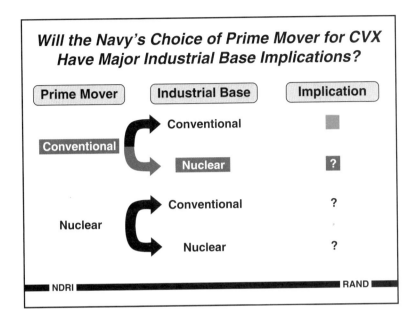

To summarize, our analysis suggests there would be no negative effects on the conventional prime mover industrial base if CVX uses either gas turbines or diesel. Both types of engines are currently widely used in marine applications, and sufficient industrial base capacity exists to easily absorb any demands from the CVX class of aircraft carriers. The only potential effect is the time—not the technology—to marinize and militarize the LM6000 for the CVX since it currently has no maritime users. However, since the LM2500 has gone through these modifications and the LM2500+ will soon be installed in cruise ships, the risks associated with marinizing and militarizing the LM6000 seem slight.[14]

Although not a propulsion system industrial base issue, a conventionally powered CVX might offer some benefits through increased competition for the construction of the ship. We did not examine

---

[14]Of course, there would be a significant integration and test effort associated with using a gas turbine on the CVX. The sizing and locating of the intake and exhaust ducting are not trivial tasks, especially when attempting to integrate the ducting into the existing (i.e., *Nimitz*-class) hull structure. We do not, however, view this problem as an industrial base issue.

this issue as part of our research. However, the Navy and the Center for Naval Analysis have considered the potential effects of competition for a non-nuclear CVX.[15]

We now consider the second case of the effect on the nuclear industrial base if the decision is made to use a conventional prime mover for the CVX class of carriers.

---

[15]See Perrin, 1998.

Even If CVX Were Conventional, Demand for Nuclear Components Will Hold or Rise

Components	Demand	Reason
Heavy equipment	Steady	Two NSSNs/year
Reactor cores Control rods Main coolant pumps Instrumentation and control Valves and auxiliaries	Steady or increasing	CVN RCOHs Trident refueling Two NSSNs/year

NDRI — RAND

The vendors that support the nuclear industrial base would certainly suffer to some degree if CVX uses a conventional prime mover. The loss of any market in an area where U.S. Navy business has been declining and where limited commercial applications exist causes concern about the vitality of the industrial base. However, the Navy nuclear program should experience some growth with the advent of the new class of attack submarines (NSSN), with the continuation of the *Los Angeles*–class submarine refuelings, and with the inauguration of the *Trident*- and *Nimitz*-class carrier midlife refueling programs.

BWXT Nuclear Equipment Division (NED) is the sole vendor that supplies the heavy-equipment components to the submarine and carrier programs. Since the heavy-equipment components are not replaced during the midlife refuelings, BWXT has only the new construction market. Also, because the Navy has always maintained a spare ship set of nuclear components for the aircraft carrier program, BWXT had little or no workload for the last of the *Nimitz* class, the CVN 77. Therefore, it has been sizing its industrial capability for some time to meet just the new submarine construction programs.[16] However, the NSSN program should provide a constant workload to BWXT for the next few years with an increase in workload around fiscal year 2004 to support the plan for two NSSNs (versus one) per year. In summary, we believe BWXT would not experience any major problems if CVX were non-nuclear.

Most of the other vendors that provide nuclear components also work in new construction and refueling. Since the *Nimitz*-class carriers have just started their midlife refueling complex overhauls (RCOHs) and the *Trident* submarines are beginning their midlife refuelings, these vendors have a new "after-market" workload. This increasing work, even with the loss of new carrier construction work, should maintain the vitality of the nuclear industrial base vendors.

In summary, we believe that a conventionally powered CVX will not cause the loss of skills or capabilities in the nuclear industrial base because of the steady or increasing workload associated with other Navy nuclear programs. There will, however, be some cost effect on those other programs if CVX is non-nuclear. We address this issue next.

---

[16]The downsizing of BWXT and its implications will be discussed later in this documented briefing.

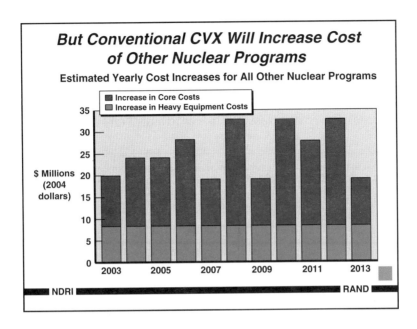

We expect the Navy submarine and RCOH programs to experience some increase in the cost of the components from vendors that supply the new construction of a nuclear aircraft carrier. This increase is primarily due to diseconomies of scale associated with lower workloads and to reallocation of overhead costs. We concentrated on the two most costly components in the ship sets for submarine new construction and refueling and for RCOHs. For example, an NSSN nuclear ship set is estimated to cost more than $400 million, about one-third of which is for the heavy equipment and the cores. Cores for a carrier RCOH and a Trident refueling cost more than $100 million and $75 million, respectively. Besides supplying the two most expensive items, the core and heavy-equipment vendors have no markets or products other than Navy nuclear components.

This chart shows the annual increase in cost for NSSN's heavy equipment and for the NSSN, carrier RCOH, and Trident refuelings if CVX is non-nuclear. The heavy-equipment cost increase is approximately $8 million per year,[17] while the increase in core costs ranges between $10 million and $25 million per year (the year-to-year variations are due to the different program requirements each year).

While some of this increase is due to production inefficiencies associated with lower workloads, much is due to a reallocation of overhead costs. The cost increases cause us to assign an orange box.

---

[17]The cost increases for the heavy equipment are the net increases after subtracting the costs associated with the purchase of new production equipment for carrier ship sets.

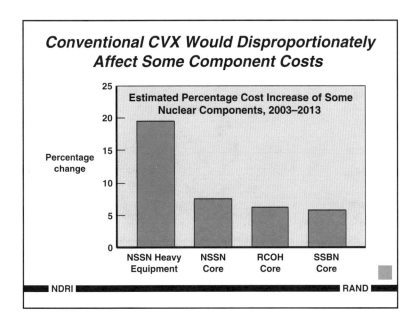

This chart illustrates the average percentage increase in the cost of the heavy equipment and the cores for the other Navy nuclear programs. The heavy equipment for NSSNs has a larger percentage increase than the cores since the heavy equipment is not replaced during refuelings. The core manufacturer has three programs (new NSSN production, *Nimitz*-class RCOHs, and nuclear-powered ballistic missile submarine (SSBN) refuelings) over which to spread increases versus only one program for the heavy-equipment manufacturer (new NSSN production). The cost increase is 5 to 10 percent of the cost of all the nuclear components for an NSSN or an RCOH and less than three percent of the total cost of an NSSN.

This increase in the cost to other Navy nuclear programs results in an orange box for the nuclear industrial base if CVX uses either a gas turbine or a diesel engine for the prime mover.

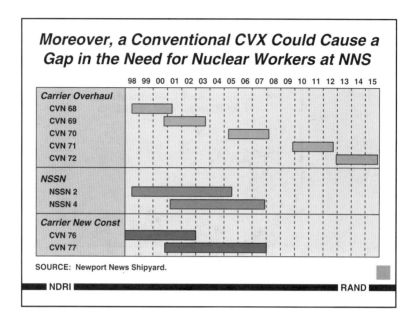

In addition to an effect on the nuclear vendors, a non-nuclear CVX would also have an effect at NNS, the shipyard that performs the RCOHs and is involved in the construction of the NSSNs. This chart shows the nuclear-related work that NNS expects if CVX is non-nuclear (the NSSNs are the two of the first four that NNS will deliver and, therefore, will require nuclear-related work).

Discussions with NNS suggest that the shipyard is concerned with the gap in the demand for nuclear workers in the fiscal year 2008 to 2010 timeframe. That gap would normally be filled with work for a nuclear CVX. The gap is somewhat mitigated by an increase in the NSSN deliveries for NNS from one every other year to one every year, starting in fiscal year 2006. However, if the projected increase in NSSN production is not realized, the full force of a non-nuclear CVX will cause a significant drop in nuclear work at NNS over the time period mentioned.[18]

---

[18] There is also the issue of whether NNS has the expertise, resources, and workforce to build a conventionally powered aircraft carrier. Discussions with NNS suggest that this would not be a problem. It has recently constructed diesel ships (its commercial Double Eagle tankers) and has designed a gas turbine frigate for the international

If a non-nuclear CVX is built at NNS, then non-nuclear work associated with the propulsion system will create a demand for labor at the shipyard. However, even if the same nuclear workers would be involved in the non-nuclear work, there will likely be some costs associated with keeping the nuclear workers qualified and trained. Time constraints did not allow us to delve deeper into this potential increase in cost. The fact that the period of concern is a decade into the future provides some time to manage the workforce. Regardless of the actions taken, a non-nuclear CVX will cause greater fluctuations in the demand for nuclear skills and will probably result in some increase in cost to the other nuclear programs at NNS. Thus, the orange box.

---

market. It has also performed overhauls and conversions of conventionally powered ships. The consensus states that it is much easier to transition from nuclear ships to conventional ships versus transitioning in the opposite direction.

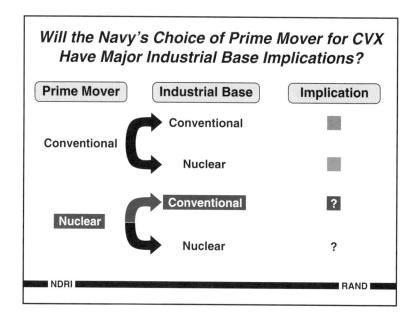

Because of the increase in the cost to other Navy nuclear programs associated both with the vendors and the shipyard, the nuclear industrial base implications of a non-nuclear CVX are assigned an orange box.

The effect on the conventional propulsion industrial base if CVX has a nuclear prime mover will now be discussed.

> **The Navy Might Have to Deal with a Diesel Monopolist**
>
> - Wärtsilä will likely close the Indiana facility without additional Navy business
> - Coltec Industries would be the domestic medium-speed diesel monopolist, whose primary customer is the Navy
> - But robust international diesel market would exist, which can be tapped if "Buy America" rules don't apply

While a variety of conventional prime mover approaches are perfectly feasible from an industrial base perspective, that does not imply that there aren't industrial base implications for the CVX prime mover choice.

For example, because of the loss of the LPD-17 propulsion system award, Wärtsilä indicated that its Mount Vernon Indiana medium-speed diesel plant is currently operating unprofitably and it plans to close it when its lease expires in about four years if it does not receive additional Navy diesel business. If Wärtsilä closes that facility, Coltec would become the domestic medium-speed diesel engine monopolist with its Fairbanks Morse Engine plant in Beloit, WI.

The Navy may or may not be troubled by such a development. Certainly, the Navy deals with monopolists in other areas—namely, sole-source nuclear vendors and the one builder (NNS) of aircraft carriers for the past 35 years. Although the Navy will need medium-speed diesel engines for other ships in the future, there is a robust international marine diesel industry that is driven largely by commercial ships that have been converted from oil-fired steam propulsion. As noted, the Navy CVX business would have little effect on the economics of this industry. The Navy could continue to obtain diesel engines at a reasonable price if it could at least credibly threaten to purchase them internationally rather than from Coltec.

The severity of this problem is directly related to the rigidity of "Buy American" policies that might affect Navy diesel purchases.[19] A green box is assigned.

---

[19] Title 41, United States Code, Section 10a, notes that American-made materials are to be used on federal projects "unless the head of the department or independent establishment concerned shall determine it to be inconsistent with public interest, or the cost to be unreasonable."

The Gas Turbine Industrial Base Is Not Affected by Nuclear CVX

- GE has a robust commercial and naval surface combatant market

- But a non-LM6000 CVX may raise development costs for other Navy surface ships

NDRI   RAND

If the decision is made to use nuclear power for the CVX, we do not envision an effect on the gas turbine engine industrial base. General Electric and other turbine manufacturers have very robust markets for both naval surface combatants and commercial ships. The loss of a prospective workload from CVX would scarcely be felt. The LM6000 engine is currently a candidate for the DD-21 program. Unless there are commercial shipping demands for this engine, the DD-21 program may be forced to incur the cost of marinizing and militarizing the LM6000 if the CVX is nuclear.

The extensive availability of gas turbines and the robust market cause us to assign a green box.

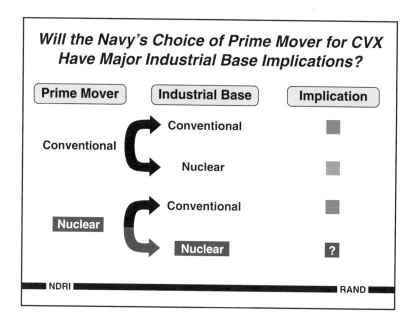

In all, we see no significant effect of a nuclear CVX on the conventional prime mover industrial base. Therefore, the green box is assigned.

We now turn to the last question, the potential effect on the nuclear industrial base if CVX uses a nuclear reactor for the prime mover.

A Nuclear CVX Would Have a Positive Effect on the Nuclear Industrial Base

For all vendors
- A nuclear CVX would perpetuate or increase demand
  - Should provide greater production efficiencies
  - But requires some buildup in production resources

For sole heavy-equipment manufacturer
- Nuclear CVX creates challenges to build up workforce, facilities, and production capabilities

NDRI — RAND

A nuclear CVX would, in general, benefit the nuclear industrial base. The workload for the vendors that support new carrier construction would either remain steady (for those with a CVN 77 workload) or increase (for the heavy-equipment manufacturer). This steady or increased workload should result in production efficiencies and reduced cost to Navy nuclear programs, compared with the costs if CVX were non-nuclear. However, an increased workload may require a buildup in production capability for some vendors. This is especially true for the heavy-equipment manufacturer BWXT.

> ### *A Nuclear CVX Would Challenge the Sole Remaining Heavy-Equipment Supplier*
>
> BWX Technologies/Nuclear Equipment Division must
> - Expand/rebuild workforce
> - Modernize the facility
> - Produce first-of-a-class design on schedule
> - Deliver components on time
>   - *To sustain a carrier force structure of 12*
>
> NDRI   RAND

With no CVN 77 work and fewer new attack submarines under construction, BWXT has been reducing its workforce and holding down overhead costs by closing facilities and selling off some of its capital equipment. A nuclear CVX would reverse that trend. The company would have to hire and train new workers, purchase new equipment, and reopen and modernize facilities. This in and of itself is a challenge when working on a tight schedule. However, a CVX nuclear reactor would be a new design, which may lead to some unexpected delays in the production schedule. Most important, the nuclear ship set is needed early in the construction cycle of an aircraft carrier and a delay in delivery of the heavy-equipment components to NNS may result in a delay in the delivery of this ship to the Navy. CVX 78 is scheduled to replace the nuclear ship *Enterprise* in 2013, after 52 years of service. If CVX 78 is late, the carrier force structure could drop below the 12 ships needed to maintain theater coverage. We will elaborate on the potential problems at BWXT in the next several charts. An orange box is assigned.

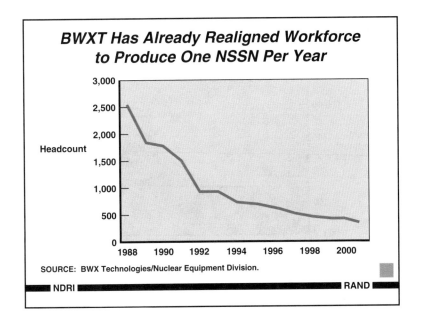

This chart shows the dramatic decrease in the BWXT workforce over the last decade. This decrease is due to the completion of carrier work for the CVN 76[20] and to the decline in new attack and strategic submarine production. To hold down overhead costs and, therefore, the cost of submarine ship sets, BWXT has been sizing its workforce to meet the demand for one NSSN ship set a year.

If CVX is nuclear, this decline will have to be reversed, thus the orange box.

---

[20]As discussed previously, CVN 77 will use the spare ship set maintained in case of a problem in producing and delivering nuclear components. Therefore, there are currently no spares for the construction of CVN 68–class carriers or for a follow-on ship that would use the A4W reactor.

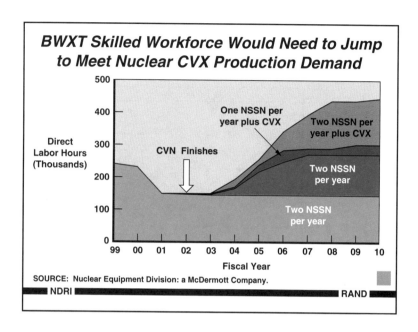

This chart shows how the demand for direct labor at BWXT would increase with a nuclear CVX and component fabrication for a second NSSN starting in fiscal year 2004. In conjunction with the workload for a second NSSN, a nuclear CVX would result in an almost tripling of the BWXT workforce in a five-year period. BWXT would have to hire and train more than 200 new direct laborers to meet this increased workload.

In recognition of the need to rapidly expand BWXT's workforce, we assign an orange box.

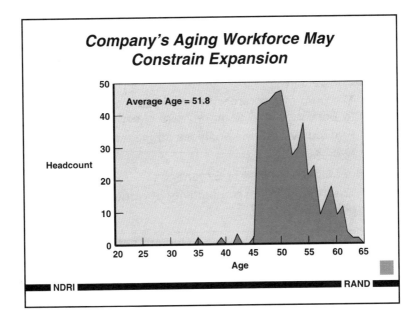

Such a large expansion in a short period of time would be difficult under most circumstances. BWXT faces an even tougher challenge given the demographics of its current workforce. Because of union rules, it has been forced to release workers during its downsizing on the basis of seniority. This has resulted in a highly experienced, but aging, workforce. The distribution of the ages in 1996 of the remaining workers at BWXT is shown in the above chart. In the two years since these data were provided, the average age of the workforce has topped 53 years and the whole curve has moved to the right. Without special provisions to retain the senior workforce, BWXT may lose workers to retirement almost as quickly as it hires new ones. The problem becomes more critical in that the experienced workers are necessary to mentor and train the new hires. Thus, an orange box is assigned.

> **BWX Workload Analysis:**
> **Key Concepts**
>
> **Mentoring Ratio**
> - The ratio of workers with 0 to 5 years of experience to workers with 5 or more years of experience
> - Previous workforce analyses suggest a ratio of at most 2:1
>
> **Productivity Gains Due to Experience**
> - The change in the productivity of workers as they gain experience
>
> ---
>
> **The interaction of these two factors determines the feasibility of meeting a set schedule**
>
> NDRI   RAND

To help understand the potential delays that BWXT may face in the scheduling of the production of the heavy-equipment components for a nuclear CVX, we built a model of its workforce expansion. There are two primary variables in the model. One is the mentoring ratio (MR)—the ratio of new hires to experienced workers. A way to think of the mentoring effect is that a core of skilled workers is necessary to effectively rebuild the workforce. Our experience with similar analyses at NNS and Electric Boat suggests each experienced worker can mentor at most two new hires.

The second variable is how quickly new hires can become fully productive. Again, our analysis of NNS suggests that new hires take up to five years to become fully productive. The values for these two variables determine how quickly BWXT can reconstitute a workforce and, therefore, the likelihood of meeting the production schedule for the heavy equipment of a nuclear CVX.

If the time for new hires to become fully productive is held constant, increasing the MR (experienced workers can mentor more new hires) will result in a faster buildup of the workforce. Also, if the MR is held constant, increases in the time for new hires to become fully productive will slow down the workforce buildup. Inherent in these two statements is our assumption that these two variables are independent. That is, we assume that changes in MRs do not affect the time it takes for new hires to become fully productive.

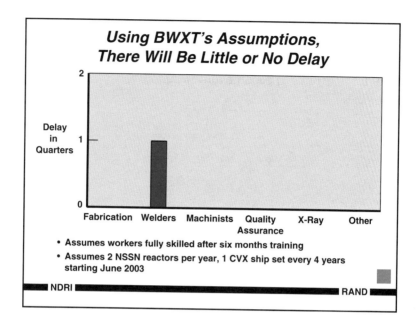

In their workforce expansion plans, BWXT assumed that new workers would become fully productive in six months. They did not use a mentoring ratio in their planning calculations. Using their assumption of new workers being fully productive in six months and a mentoring ratio of two new hires for every experienced worker, our model suggests that BWXT could meet the production schedule for the new CVX reactor components. We, therefore, assign a green box if BWXT's workforce expansion assumptions are correct.

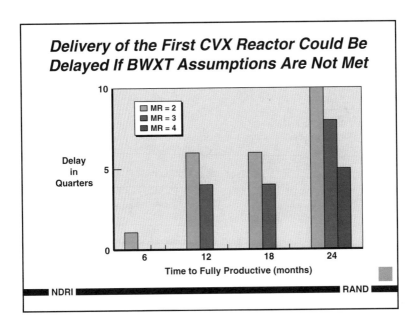

We used our model to examine the effect on the delivery of the first CVX reactor if BWXT's assumptions were overly optimistic. In the chart, we show the estimated delay in quarters for different values of the mentoring ratio and changes in the time to get new workers fully productive. Here, as before, we assume that starting in June of 2003 BWXT will build two NSSN ship sets per year and one CVX ship set every four years.

As we discussed in the previous chart, a productivity time of six months and an MR of two to one will result in little or no delay in the delivery of the components. If new workers become fully productive in 12 or 18 months, then an MR of four to one is needed to meet the delivery schedule. If the time to get new workers fully productive takes 24 months, which is still less than the times experienced by various shipyards we've examined, even higher MRs will result in significant delays in the delivery of the components. With a 24-month productivity time, an MR of two to one results in an estimated delay of approximately two to three years.[21] Higher MRs will reduce the delay, but even assuming an MR of four to one results in estimated delays of over one year. Because of the potential delay in the delivery of the CVX reactor components, we assign an orange box.

---

[21] In all cases, the need to hire and train new welders is the primary driver of the delays. Other skills also result in delays in delivery, but welding is the most critical skill for BWXT.

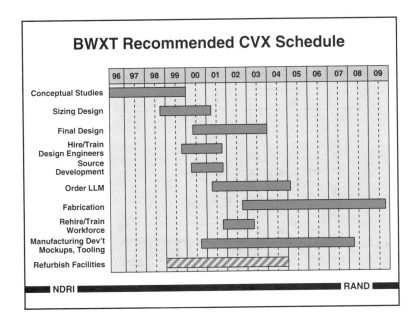

The potential problems in reconstituting the BWXT workforce are magnified when put into perspective of other critical tasks in the process of producing the heavy-equipment components for a nuclear CVX. This chart shows the current recommended schedule for CVX design and production. The first four tasks in the chart relate to the design of the new reactor for CVX. Because it is the first of a class, there is always the potential for slippage in the design schedule, especially when new design engineers must be hired and trained.[22] The fabrication of the components cannot begin until the reactor has almost reached the final design point.

The potentially larger problem lies in the refurbishing of production facilities and the purchase of new production equipment. We have shaded the bar for this task differently because some of the new equipment has lead times of up to two years. Since the equipment must be available for the training of new employees, any delay in ordering or receiving the equipment can cause further delays in re-

---

[22]The laboratories that support the Navy nuclear programs will accomplish the actual design of the new reactor. BWXT uses design engineers to translate the reactor design to efficient production operations.

constituting the workforce. The new equipment that is needed for the production of one NSSN per year plus a CVX includes the following:

Equipment	Cost ($M of 1998)
Large machine tools	23.7
Welding equipment	3.0
Robotics	1.8
Small machine tools	5.8
General equipment/facility activities	14.0
Foundations for equipment	4.0
Total	52.3

To put the schedule in perspective, NNS needs components by mid-2008. Under normal circumstances (i.e., with an existing reactor and an experienced workforce), the lead time for components is approximately seven years. This includes two years for the necessary forgings. With a minimum six-month period to hire and train the new workforce and a two-year period for purchase and delivery of the necessary new production equipment, there seems to be no slack in the recommended schedule. The long–lead time equipment must be ordered no later than January 1999.

Both BWXT and NAVSEA's Nuclear Propulsion Directorate are aware of the potential problem and are trying to manage the reconstitution of the heavy-equipment production capability. However, it is difficult to take the necessary actions to reduce the schedule risk until a final decision is made on the propulsion system for CVX.

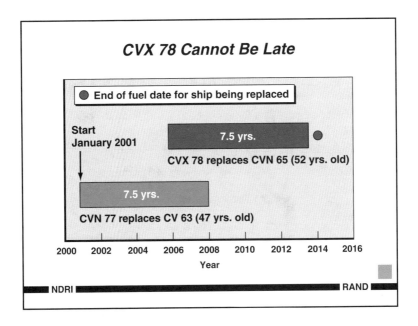

The importance of timely delivery of nuclear components to NNS is magnified when considering the effect of a delay in delivery of the first CVX to the Navy. As previously mentioned, the first CVX is scheduled to replace the *Enterprise* in the fleet. CVX 78 must be in service by 2013 to ensure the timely replacement of *Enterprise*, after 52 years of service. Any delay would cause problems in providing adequate carrier coverage of the major theaters of operation.

Because of this risk in the production schedule for the heavy-equipment components, we assign an orange box to the effect on the nuclear industrial base if CVX uses nuclear propulsion.

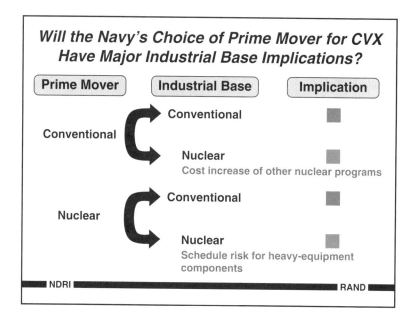

This chart summarizes our findings about the effect of the CVX propulsion decision on the conventional and nuclear propulsion industrial bases. Regardless of whether CVX is nuclear or non-nuclear, we find no significant effect on the diesel or gas turbine engine industrial bases. The manufacturers in those areas have very robust markets in other naval ships and/or in the commercial sector. A presence or absence of demand from a conventionally powered CVX would scarcely be felt.

However, there are some concerns about the nuclear industrial base whether CVX is nuclear or non-nuclear. If CVX is conventional, the cost of components for other Navy nuclear programs would increase. Those other programs would bear more overhead, for instance. If CVX is nuclear, there is a potential schedule problem with the delivery of the heavy-equipment components. This latter problem suggests that the decision on the propulsion system for CVX must be made very soon and that the reconstitution of the production capability at BWXT must be closely managed.

Actions Can Be Taken to Reduce the Risk In the Schedule for the Heavy Components

- **Provide BWXT AP funding earlier than currently planned**
- **Work with NNS to change the delivery requirement for the heavy equipment**
- **Use the workforce expansion data to closely monitor BWXT**

NDRI — RAND

While we have identified schedule risk associated with the nuclear industrial base if CVX is nuclear, we also believe multiple strategies exist to decrease this risk. First, by providing advanced procurement (AP) funds earlier than currently planned, the Navy could encourage BWXT to begin facility upgrades and hiring earlier than the current plan. If delivery dates do slip, the ship construction sequence could be adjusted. Finally, in the course of this research we developed manpower growth curves by skill, which could be used to ensure that the BWXT workforce expands at the rate necessary to maintain the schedule.

Appendix A
PRIME MOVER ATTRIBUTES

Prime Mover Attributes

Type	Status	Advantages	Disadvantages
Oil-Fired Steam	Last Navy ship is LHD, delivery 2001; 3 CVs in service	• Fuel efficient @ cruise speed • High output power	• Large, heavy plant • High manning reqts • Corrosive exhaust gas
Gas Turbine	All Navy surface combatants built since mid-70s	• Light-weight, compact • Fast startup, response	• Limited output power • Large air intakes/exhausts • Low efficiency @ partial pwr • High infrared signature • Corrosive exhaust gas
Diesel, Medium Speed	Navy fleet auxiliaries, amphibious ships	• Fuel efficient @ all loads • Low fuel costs	• Limited output power • Corrosive exhaust gas • High pollution
Nuclear Steam	All Navy subs, carriers built in the last 30 years	• Endurance • No air intakes/exhausts • High output power	• High initial cost • Disposal following overhaul, retirement

NDRI   RAND

The venerable oil-fired steam engine provides high power output per installation, but its large size and manning requirements have caused it to be supplanted by the other types. This is true for both military and commercial applications for foreign and domestic operators.

The gas turbine, with its light weight, fast response, and agility has been found to be ideal for the smaller surface combatants, where intake and exhaust ducting challenges are not severe. Some high-speed ferry boats and pleasure craft have also chosen gas turbines for propulsion.

The high fuel efficiency and low fuel cost of diesel engines have been found appropriate for those ships where a large fraction of their operation is at constant cruise speeds, such as tankers and freighters. This is true for both Navy and commercial operators.

The endurance attribute of the nuclear propulsion system makes it ideal for the mission of Navy fleet ballistic missile (FBM) submarines, attack submarines, and aircraft carriers. Freedom from intakes and exhausts saves space and enhances the ability to defend against airborne chemical, biological, and radiological threats.

Appendix B
ENGINE EXHAUST AIR POLLUTANT EMISSIONS

Air pollution resulting from the exhaust emissions of marine engines is becoming an increasingly important concern for the world's navy and marine interests. Estimates of air pollution from marine sources in West Coast cities range from 4 to 10 percent of the total, and these proportions are expected to increase as land sources come under increasingly stringent control.

The principal pollutants of concern from marine engines are (1) oxides of sulfur or SO_x, (2) particulate matter or smoke, (3) hydrocarbons (unburned portions of the fuel), and (4) oxides of nitrogen or NO_x. The first two are best controlled by the selection of fuel. Fuel with low sulfur and ash content lead to low SO_x and smoke in exhaust.[1] The latter two are a consequence of the combustion parameters, such as cylinder pressure, air content, fuel injection timing, etc., and are best controlled by making engineering changes in the way engines are designed and operated. Carbon dioxide and carbon monoxide are also of concern. Carbon dioxide is a direct combustion product that is also a greenhouse gas, leading possibly to higher temperatures on the surface of the Earth. Carbon dioxide emissions are directly proportional to the fuel efficiency or "gas mileage" of an engine. Carbon monoxide results from incomplete combustion and can be a concern to the sailors operating the engines.

---

[1] Combustion of lubricating oil is also a contributor to smoke emissions from marine diesels. The burst of heavy smoke observed on highways as diesel trucks accelerate or change gears is due to a burst of lubricating oil consumption.

In all fossil-fueled engines, SO_x emissions are directly proportional to the sulfur content of the fuel and to the fuel efficiency of the engine. Gas turbines require refined fuel that has a low sulfur content. Navy diesels burn the same low-sulfur fuel and achieve even lower SO_x emissions because of their superior fuel efficiency. In commercial practice, the ability of diesel engines to burn a variety of fuels, including inexpensive heavy fuel oil with high sulfur levels, is one of their advantages. Thus, commercial diesel-powered ships normally emit more SO_x than a ship with equivalent gas turbine power.

Table B.1 compares typical exhaust emissions of a large gas turbine and large diesel engine. These numbers illustrate typical differences. Any particular engine can vary from these values because of operating conditions, state of maintenance, or other variables.

Table B.1

Average Pollutant Emissions from Marine Diesels and Gas Turbines Without Supplemental Emission Controls

Source	Diesels[a]				Gas Turbines[a]			
	NO_x	CO	HC	PM	NO_x	CO	HC	PM
NAVSEA[b]	15	0.5	0.4	0.2	8	0.1	0.02	nil
USCG[c]	9	0.6	0.1	—	6	0.1	0.1	—
USCG[c]	12–20	1.4–1.9	0.1–0.5	0.1	—	—	—	—
NAVSEA[d]	—	—	—	—	3.2	0.5	0.1	0.1
GE[e]	—	—	—	—	3.5	0.1	0.01	nil
Wärtsilä[f]	10	—	—	—	—	—	—	—
Risø Natnl. Lab., Denmark[g]	12–17	1.6	0.5	—				
Average	13	1.1	0.3	0.2	5	0.2	0.06	nil

[a]All data are in grams per kilowatt-hour (gm/kw-hr).
[b]Exhaust emission Generation Mechanisms, JJMA to NAVSEA, 8/18/98.
[c]Shipboard Engine Emission Underway Testing, USCG by Allied Marine Services, 8/18/98.
[d]Navy Engine Exhaust Emission & Database Web Tool presentation, p. 10, 8/18/98.
[e]Gas Turbine Emission Technology Overview, GE, 8/18/98.
[f]Technology for Diesel Exhaust Emission Reduction, Wärtsilä NSD, 8/18/98.
[g]Niels A. Kilde and Lene Sørensen, Risø National Laboratory, Systems Analysis Department, Roskilde, Denmark.

CONTROL OF MARINE DIESEL AIR POLLUTANT EMISSIONS

The focus of the United States Environmental Protection Agency's (EPA's) current regulatory work on marine engine emissions is on piston engines. These include gasoline or "spark-ignition" engines and diesel or "compression-ignition" engines.[2] The EPA's focus arises from two main reasons: (1) about 95 percent of the world's commercial and recreational vessels are powered by piston engines, and (2) except for carbon dioxide, piston engines inherently emit greater amounts of pollutants (per gallon of fuel or per kilowatt-hour) than gas turbine engines. In 1996, the EPA began to concentrate on a rule that would be applicable to large marine diesel engines. This rulemaking effort is compounded by recent international rules issued for marine engines by the International Maritime Organization (IMO)[3] and further by the forthcoming Law of the Sea convention. If the United States ratifies this convention, the United States would cede to the IMO certain environmental regulatory power over ships in international trade that call on U.S. ports. No rule has yet been proposed, but, at present, the EPA is envisioning a rule that would apply to diesel engines on U.S. flagged ships or vessels in domestic commerce. The envisioned rule is necessarily complex because of the large variety of marine propulsion systems and the ways in which they are employed and maintained. Issues such as the engine duty cycle, ports in which they are operated, maintenance plans and execution, and fuel choices available to the operator all affect what can and should be done to reduce emissions. In general, the EPA envisions a rule that would reduce the permitted NO_x emissions below that in the IMO rule[4] for large marine diesels.

---

[2]Since 1992, the EPA has issued a rule applicable to outboard and personal watercraft engines that is expected to result in a 75 percent reduction in the emission of unburned hydrocarbons from these engines over the nine-year phase-in period of the rule. Currently, the agency has also elected not to regulate emissions from stern-drive engines.

[3]MARPOL Annex VI, Regulation 13 would apply to marine diesel engines. This annex is of recent origin and is not yet in force because the necessary number of countries has not ratified it.

[4]When in force, the IMO rule will limit diesel NO_x emissions from large marine diesel engines to a sliding scale depending on the maximum rated engine speed. Very-low-speed engines, such as those used in direct-drive propulsion systems in oil tankers, will be permitted to emit up to about 15 gm/kw-hr. Medium-speed engines typically used for diesel electric propulsion systems in cruise ships will be limited to about 12

In response to the forthcoming rules on diesel engine emissions, diesel manufacturers have developed several solutions.[5] Careful design of the engine and adjustment to the combustion conditions can reduce NO_x emissions to within the proposed IMO rule with no further action. Further reductions of NO_x to levels close to those achieved by current gas turbines and within the notional EPA rule can be realized by injecting water into the diesel cylinders along with the fuel.[6] Reduction of NO_x emissions to levels approaching the best of the low-emission gas turbines can be achieved by using selective catalytic reduction (SCR) systems, which decompose the NO_x in the engine exhaust to nitrogen and oxygen. This is analogous to the catalytic exhaust gas purifiers used in modern automobiles. SCR machines are technically proven but are very large and would occupy space that could be used for other purposes. Their size may prohibit their use in existing ships. Efforts are under way to reduce the size.

Whether a diesel exhaust emission rule would be applicable to U.S. military diesel-powered vessels is not clear, but, currently, the EPA is considering that it would apply to vessels used by the military in a manner comparable to civilian vessels. Tugboats and shipyard watercraft are examples. Warships would be exempt. The Navy has long used only low-sulfur, low-ash fuel for all of its fossil fuel engines, so no further improvement with regard to SO_x emissions is anticipated. The Naval Sea Systems Command is drafting a Navy rule that would require future Navy diesels to incorporate the best available emission control design features.

---

gm/kw-hr. High-speed diesel engines, such as those commonly used to power ship's electric power generators, will be limited to about 10 gm/kw-hr. The IMO rule will also limit fuel sulfur content to 4.5 percent for worldwide service and to 1.5 percent for service in sulfur-control areas. At present, the only sulfur-control area is the Baltic Sea.

[5]The following is courtesy of Wärtsilä NSD.

[6]The precise timing of fuel and water injections is important. In practice, water injection begins before fuel injection begins and ends just after fuel injection begins. About 0.5 gallons of pure water is required for each gallon of fuel.

CONTROL OF MARINE GAS TURBINE AIR POLLUTANT EMISSIONS

Marine gas turbine engines power many new Navy surface warships and have been considered for CVX. Gas turbines offer many advantages over other fossil fuel alternatives, including less space and weight for the propulsion system and reduced maintenance and, as shown in Table B.1, less pollution, except for CO_2. The Naval Sea Systems Command has calculated the emissions of NO_x and other pollutants from a gas-turbine-powered aircraft carrier.[7] Such a ship operating for about 3,100 hours in a year would emit about 1,000 tons of NO_x from her main engines and auxiliaries. Emissions vary ±20 percent depending on the specific engines chosen and the ship's assumed operating pace. A diesel-powered aircraft carrier would emit more NO_x (and other pollutants except CO_2) than a gas-turbine-powered aircraft carrier, but diesel propulsion is not being considered for carriers because of performance limitations that make diesel systems unsuitable.

The annual NO_x emission for one gas-turbine-powered aircraft carrier would be a very small fraction of the 11 million tons of NO_x emissions from all the world's ships or the 1.5 million tons emitted by all of the world's military ships. Compared with just the U.S. Navy ships, NO_x emissions from a single gas-turbine-powered aircraft carrier would increase the current (calendar year 1997) Navy ship emissions from about 22,000 tons to about 23,000 tons, or 5 percent. Assuming the non-carrier elements of the fleet remained the same, a fleet of 10 such ships would be responsible for about one-third of all the Navy's NO_x emissions.[8] In terms of per-capita emissions, a gas-turbine aircraft carrier would emit about 0.17 tons per crewman compared with just under 0.10 tons per capita for the average person living in the United States.[9]

---

[7] Personal communication between D. Rushworth (MSCL, Inc.) and M. Osborne, (NAVSEA 03Z), July 23, 1998.

[8] From Navy Exhaust Emission Modeling, a presentation given at the Marine Engine Exhaust Emission Workshop held at the Naval Sea Systems Command, August 18, 1998.

[9] From EPA NO_x emission data for 1992 and U.S. Bureau of the Census data for 1992.

The EPA has advised that there is currently no plan to consider regulation of marine gas turbine engines.[10] However, gas turbine engines are regulated when used in land applications, such as in natural gas pipeline compressors. For this reason, General Electric, a manufacturer of gas turbines, has developed two means to reduce emissions. The simplest of these, from an engine design standpoint, is water injection.[11] By injecting about 0.9 gallons of pure water into the engine for every gallon of fuel, NO_x emissions are reduced about sixfold, but CO and hydrocarbon emissions are increased about threefold. So much pure water is needed that long-term use of the technique in a warship is impractical. General Electric has also developed improved combustors that promise to reduce NO_x emissions by a factor of 10 or more with little or no degradation in CO and hydrocarbon emissions. The improved combustors are presently available only for natural-gas-fueled gas turbines but are expected soon for marine gas turbines. The Navy is also working on comparable technology in anticipation of future regulation.[12]

AIR POLLUTANT EMISSIONS FROM NUCLEAR-POWERED AIRCRAFT CARRIERS

Nuclear-powered aircraft carriers also emit air pollutants, but not from their nuclear power plants. Such ships are equipped with small diesel-powered electric generators to provide electric power when the nuclear reactor(s) are not operating. NAVSEA has estimated that all such diesel generators in all nuclear-powered ships and submarines emitted less than 170 tons of NO_x in calendar year 1997. This compares with nearly 28,000 tons of NO_x emitted from all Navy ships and craft (piston driven, gas turbine, fossil-fueled steam, and nuclear powered) during the same year. Although nuclear aircraft carrier designers may select modern low-emission diesels or even emission-controlled gas turbines for their backup electric power sys-

---

[10] Personal communication between D. Rushworth (MSCL, Inc.) and J. M. Revelt, EPA, August 18, 1998.

[11] "Gas Turbine Emissions Technology Overview," David L. Luck, General Electric Corporation, August 18, 1998.

[12] From the Navy's Ship Environmental Research and Development Program (SERDP) project titled "Reduction of NO_x Emissions from Gas-Turbine Power Plants."

tem, the choice will have virtually no effect on the Navy's overall fossil-fuel exhaust emissions.

CONCLUSIONS

Air pollutant emissions from aircraft carriers will be affected by the choice of propulsion for CVX. A single gas-turbine-powered aircraft carrier would increase current Navy-fleet air pollutant emissions by about 5 percent and a fleet of 10 such ships could account for about one-third of all of the Navy-fleet air pollutant emissions. The existing U.S. and European industrial base has developed emission-reduction technology that is presently available or will soon be available to reduce engine emissions severalfold. The emission reduction technology will enable future engines to comply with anticipated future emission-reduction regulations and Navy standards.

Air pollutant emissions from the auxiliary fossil-fuel engines in nuclear aircraft carriers are an insignificant fraction of all Navy-fleet air pollutant emissions and can be further reduced by employing modern engines and emission-reduction technology.

Appendix C

ARRANGEMENT CONSIDERATIONS FOR THE AIR INTAKES AND EXHAUSTS FOR FOSSIL-FUELED AIRCRAFT CARRIERS

The Navy is evaluating a variety of fossil-fuel-powered aircraft-carrier alternatives, all based on using gas turbines, either in direct-drive or electric-drive configurations.[1] A notional direct-drive configuration would use two General Electric model LM2500+ gas turbine engines driving through a reduction gear to power each of four propeller shafts. Four additional LM2500+ gas turbines would power electric generators to provide for the ship's electric requirements. A notional electric-drive configuration would use six General Electric model LM6000 gas turbines to drive electric generators. In this configuration, the propeller shafts would be driven by electric motors that would use most, but not all, of the electricity from the LM6000 generators, leaving about 25 percent of the electric power for the ship's electric needs, or even more when high propulsion power is not required.

Gas turbine engines require large amounts of air to feed the turbine (the "intakes") and equally large exhaust gas plenums to carry away the waste gas (the "uptakes"). These plenums would occupy several thousand cubic feet of space within the upper decks of ship in areas that, in nuclear-powered aircraft carriers, are occupied by crew quarters, storerooms, weapons systems and other essential ship's systems. Of particular note, intakes and uptakes for gas turbine sys-

---

[1] Personal communication between D. Rushworth (MSCL, Inc.) and J. Dunne (NAVSEA 03Z), July 1998.

tems would occupy 400 m² of floor space on the ship's main or "hanger" deck, or about 10 percent of the total main deck area. By careful arrangement of the intakes and uptakes, the obstruction created by these structures can be reduced to no more than that otherwise occupied by one aircraft.[2] Table C.1 shows an approximate comparison of the deck areas that can be occupied by the intakes and uptakes of one particular gas turbine concept.[3]

Estimates such as those in the table suggest that gas turbine engines will significantly encumber the useful deck area of the upper decks of an aircraft carrier. However, gas turbine engines are very compact for the power they produce, compared with diesel- and nuclear-powered engines, and therefore there are space savings in the lower

Table C.1

Impact of Gas Turbine Intakes and Uptakes On the Useful Deck Area in a Notional CVX

Deck Level	CVX C3[a]				
	Intake	Uptake	Totals	Total Available Deck Area[b]	% of TADA
Main	307.5	94	401.5	~4,100	10
01	365.9	235[b]	600.9	~2,000[c]	18[d]
02		242[b]	242	—[c]	—[c]
03		108	108	~34,000	0.3
Total	615.4	679			

[a]All values are in square meters and are approximate based on estimates from a typical CVN 68 class ship.

[b]This table assumes that the deck areas in CVX are comparable to those in the CVN 68 class.

[c]Much of the 01 and 02 ship levels are for the air space of the hanger deck. Usable area on these decks is along the sides and at either end of the ship. Uptakes will probably be located in the air space and therefore will not encroach on the available deck space. Intakes on the 01 level will occupy useful space on this deck.

[d]Considers only intake area.

---

[2]Personal communication between D. Rushworth (MSCL, Inc.) and J. Dunne (NAVSEA 03Z), July 1998.

[3]Intake and uptake deck areas are from CVX Study C3 by the Naval Sea Systems Command, Code 03Z.

decks that can be used to house some of the functions displaced from upper decks. Also, careful design can place intakes and uptakes so as to minimize their adverse effects. Thus, only a careful comparison of notional designs of entire ships can provide a useful assessment of intake and uptake effects. Of note is that the Navy is considering aircraft carrier designs that have two islands because of certain operational improvements, and the choice of gas turbine propulsion would almost certainly require two islands to house the uptakes.

There are no industrial base considerations regarding these issues. Many existing Navy ships use gas turbine (or diesel) engines, and many design approaches are available to accommodate the intake and uptake issue.

Existing nuclear-powered carriers are not encumbered with significant engine intakes and uptakes in the upper decks. They have only one island, which houses ship and flight control functions, radar, communications equipment, and other equipment. The ship's power systems include small diesel electric-power generators that require very small intakes and uptakes in the upper decks.

In conclusion, intakes and uptakes will affect the arrangement of the upper decks of an aircraft carrier. Some of the effects on the upper decks are mitigated by increased available space on the lower decks. There are no industrial base concerns regarding intakes and uptakes.

BIBLIOGRAPHY

AoA, *CVX AoA Part 2, Propulsion and Electrical Machinery Plant and Ship Concept Evaluations*, Alexandria, Va.: Center for Naval Analysis, March 27, 1998.

Birkler, John, et al., *The U.S. Aircraft Carrier Industrial Base: Force Structure, Cost, Schedule, and Technology Issues for CVN 77*, MR-948-Navy/OSD, Santa Monica, Calif.: RAND, 1998.

Birkler, John, et al., *The U.S. Submarine Production Base: An Analysis of Cost, Schedule, and Risk for Selected Force Structures*, MR-456-OSD, Santa Monica, Calif.: RAND, 1994.

Borman, J. B., *The Electrical Propulsion System of the QE2: Some Aspects of the Design and Development*, Rugby, UK: Cegelec Projects Ltd., 1994.

Cegelec Projects Ltd., *A Step Ahead in Propulsion: The POD Concept*, Rugby, UK: Cegelec Projects Ltd., Kamewa Propellers, February 1998a.

Cegelec Projects Ltd., *Marine and Offshore Power and Propulsion Systems*, Rugby, UK: Cegelec Projects Ltd., Industrial Systems Group, 1998b.

Dade, Tom, *Electric Drive—Coming of Age* (briefing), Newport News, Va.: Newport News Shipbuilding, September 1997.

Davis, James C., *A Comparative Study of Various Electric Propulsion Systems and Their Impact on a Nominal Ship Design*, Cambridge Mass.: MIT, June 1987.

Diesel & Gas Turbine Publications, *Diesel & Gas Turbine Worldwide Catalog*, Brookfield, Wisc.: Diesel & Gas Turbine Publications, 1998 (see http://www.dieselpub.com).

GE Aircraft Engines, *A Proud Heritage of Power*, Evendale OH: GE Aircraft Engines, June 1996.

Goodman, Glenn W., "Three New Ships Hold the Key to Long-Term Modernization of Surface and Submarine Fleets," *Armed Forces Journal International*, March 1998.

MAN Aktiengesellshaft, *MAN Annual Report 1996/97*, Munich, Germany: MAN Aktiengesellshaft, October 1997.

Metra Corporation, *Metra Corporation Annual Report 1997*, Helsinki, Finland: Metra Corporation, March 1998.

Newport News Shipbuilding, *Future Carrier Propulsion Options*, briefing presented to RAND on July 7, 1998.

Newport News Shipbuilding & Electric Boat, *CVX Evaluation of an Electric Drive Propulsion Plant Concept*, Newport News, Va.: Newport News Shipbuilding & Electric Boat, April 1998.

Perin, David A., *Cost Comparisons for Nuclear and Gas Turbine Designs: Summary of Cost Analysis for Part 2 of the CVX AOA* (annotated briefing), Alexandria, Va.: Center for Naval Analysis, CAB 98-102, November 1998.

Rodrigues, Louis S., *Naval Ship Propulsion: Viability of New Engine Program in Question*, Washington, D.C.: GAO, NSIAD-96-107, June 1996.

Saul, Robert, *GE Marine Engines: Executive Overview* (briefing), Evendale, Ohio: GE Marine Engines, June 1998.

Schrader, Alan R., *Technical Studies of General Electric LM2500 Marine Gas Turbine Engines in GTS Adm William M. Callahan*, Washington, D.C.: George Washington University, June 1975.

Simmons, L. D., *Naval Propulsion Systems, Phase 1: Survey of Alternative Technologies*, Alexandria, Va.: IDA, P-2532, February 1991.

Sowder, Michael, *Future Carrier Propulsion Options* (briefing), NNS IR&D Task 2004-3062G, Newport News, Va.: Newport News Shipbuilding, July 1998.

Stanko, Mark T., *An Evaluation of Marine Propulsion Engines for Several Ships*, Cambridge, Mass.: MIT, May 1992.

U.S. Naval Sea Systems Command, *Propulsion and Electrical Machinery Plant & Ship Concept Evaluations, CVX-AoA Part 2*, U.S. Naval Sea Systems Command, March 1998.

Wärtsilä Diesel International, *Wärtsilä Diesel Power News 1/95 & 1/96*, Vassa, Finland: Wärtsilä Diesel International, January 1995 and January 1996.

Wärtsilä NSD Corporation, *Marine News No. 1, 1997*, Vassa, Finland: Wärtsilä NSD Corporation, June 1997.

Weiler, Carl L., *WR-21 Intercooled Recuperated Gas Turbine*, Sunnyvale, Calif.: Northrop Grumman Marine Systems, May 1998.

Whiteway, Roger N., and Thomas Vance, *The Military Effectiveness of CVX Propulsion Alternatives*, MacLean, Va.: SAIC, September 1998.